THE REALITY OF CLIMATE CHANGE

The Biggest Threat to All of Humanity

And Life Forms on Earth

BY ALASTAIR R AGUTTER

1

QUOTATION

"The refusal to change by clinging onto superficial material wealth at any cost today is a certain sentence of death tomorrow, for all future priceless life entities we have come to love and know."

~ Alastair R Agutter

THE REALITY OF CLIMATE CHANGE

BY ALASTAIR R AGUTTER

First Recorded and Published 26th September 2015.

Printed, Published and Distributed by

Create Space Independent Publishing

An Amazon Group Company

ISBN-10: 1514261359

ISBN-13: 978-1514261354

CONTENTS

QUOTATION

"For a more knowing and informed world, we are becoming forever more complicit based on the evidence presented to us, regarding the demise of our Earth we all know as home."

~ Alastair R Agutter

NOTE

Many say if there was a real just God the events on Earth would not happen, surrounding Extremism, Social Unrest, Conflict and Climate Change.

I say Divine Creation does exist through Natural Law and the Earth does not only host Human Beings, but countless millions of life specie entities that can only be described as beautiful miracles of evolution.

As only one species, the human race is issuing a sentence of death to all life we know on Earth and the planet is dying from human greed and arrogance.

At the time of writing this book, there are two specific numbers you need to remember and these are 275 and 400!

What relevance are the above numbers you may rightly ask?

FACT: 275 Is the parts per million number of Carbon Dioxide for starting all of life on Earth and in layman terms, many millions of years ago.

FACT: 400 Is the parts per million number of Carbon Dioxide today in the Earth's atmosphere and these factual measurements are courtesy of friends and colleagues at NASA (nasa.gov).

Written within this book for all, are the facts and these writings are not the whims of liberalism, or any other body that would give cause for today's Corporate Capitalist to submit a plea of justifiable denial.

The Human Race has become a menace and a threat to all life on Earth and therefore Man must change and now!

QUOTATION

"It is not for I to preach or lay blame at ones door, but to present the facts for all who have found within them humanity, so one can seek the answers for themselves, to endorse their own growing suspicions."

~ Alastair R Agutter

INTRODUCTION

In 1988, now over 25 years ago, I wrote a book on a highly evolved marine species the symphysodon, found in the River Amazon and adjoining Rivers in South America. At the time of writing the book, I clearly recall expressing my concerns in the publication's introduction, surrounding the survival of this specie and others, as a result of man's aggressive deforestation to the Amazon Rain Forest.

Now today in the 21st Century and in the year 2015, some 25 plus years later from when I first expressed concern, the Political Elite World-Wide, many who are servants to money and Capitalism, are still pondering and debating climate change as the planet continues to now die!

In 2012 and 2013 I wrote two further articles on Climate Change and the consequences, copies I forwarded to former Vice President Al Gore of the United States of America.

The articles detailed both evidence and the eventual outcomes from ignoring the reality of Climate Change and much of the findings were from personal observation of plant and marine life.

I have been further prompted to write this short book and guide for all to read, as a result of health problems towards the end of last year (2014) and this year (2015) to folk in the community, including personal friends and acquaintances.

I have also witnessed further evidence this year (2015) as a result of Climate Change, surrounding pollination of food crops, giving great cause for further grave concern.

The facts found in this book for all to read, I am sure will answer many searching questions folk are now asking, especially surrounding health problems that include respiratory and other ailments, now becoming more prevalent to millions of folk around the World, as a result of Climate Change caused from industrial pollution.

Time is now running out and we all need to change for the preservation of all life on Earth.

Sincere Best Wishes,

Alastair R Agutter

Author

QUOTATION

"The time has come where the few can no longer do all the heavy lifting. A time has now arrived where the many must carry the burden of responsibility for all life on Earth as primate custodians."

~ Alastair R Agutter

CLIMATE CHANGE TO OUR SEAS

I began fishing as a small boy and became renowned in my community as an Angler and where today I can now share the secret to my success and this was as a result of understanding the environment and conditions.

Photograph Courtesy of Author Alastair R Agutter

My angling and marine interests allowed me to spend many hours, weeks and years around shorelines or several miles out wondering sand banks and estuary silt environments. Especially when studying the saltwater pools that housed countless specie inhabitants from small vertebrates, crustaceans to small marine fish species.

Now spanning 50 plus years of knowledge and experience of such environments, I can only say I truly despair of the continued decline, but a decline now gathering significant momentum and even witnessing first hand wild marine species stressed by the variation in water conditions and temperature even on a daily basis.

On returning to England flying back from Jersey in 2004 one early summer evening, I could not help but notice the algae bloom in the English Channel and North Sea. Such a sight I found deeply disturbing for significant algae blooms of this nature meant a huge reduction in oxygen content in the Sea.

So what is the real situation to our Seas surrounding Climate Change?

Having bred some of the rarest and most difficult marine tropical fish species in captivity for over 46 years as an Aquarist and an accomplished author writing on the subject, I can shed factual light on the reality impact of climate change to our seas and estuaries.

The key factors to breeding tropical marine species successfully in captivity is to create the correct conditions and this relates to water chemistry, food and the biological filtration environment.

Our seas, estuaries and rivers are also dependent upon having these stable conditions in place for life to exist and thrive. When the conditions are not right in a scaled down aquarium environment, fish species become stressed, stop eating, catch diseases, stop reproducing and eventually die.

The tolerance levels to certain species of marine life vary. Some species are more resilient to change, where others are affected by even the slightest of fluctuations in conditions to a whole host of marine variants relating to temperature, biological bacteria content, specific hardness, degrees of hardness, specific gravity surrounding salt content in water, oxygen, Co_2, ammonia, nitrate levels etc.

When conditions change be it in aquariums, our seas, estuaries or rivers there are serious consequences as pointed out above surrounding the health of all marine life.

Photograph Courtesy of Author Alastair R Agutter – Breeding Symphysodon in Captivity

External factors to our Seas also have a significant bearing to the health of these environments, namely the Earth's atmosphere where jet stream patterns influence temperatures and conditions.

It is a known fact in decades gone by the old hands in angling and commercial fishing solely dependent on their knowledge of conditions without the aid of any modern day technology were able to predict and accurately forecast locations and times when certain fish species were plentiful or very poor in numbers.

These cycles surrounding catches were as a result of freak weather conditions where jet streams changed direction years earlier affecting the migration patterns of fish and this impacting on their preferred natural breeding habits and locations.

Such events were on rare occasions and normally as a result of a major natural disaster somewhere in the World be it a volcanic or an earthquake event.

When such Natural Disasters happened Marine fish stocks catch rate would not necessarily be affected that particular season but some years later as a result of spawning that season where new born fry were lower in numbers and survival rate far lower as a result of condition changes to the environment and so if a fish species only reaches adult size after say three years before shoaling for example. The impact is some three years later from when the event happened and the calculations of old hand anglers and commercial fisherman could relate to why fish levels and stocks for that particular species and season was down.

Other conditions that affect the marine environment today more than ever is from man-made natural disasters and with a devastating impact.

The natural cycles in nature that do change from time to time is challenging enough for the marine environment. But when man-made disasters happen in addition to natural cycles of fluctuation in the environment, such events can have a permanent outcome to the landscape affected and the specie inhabitants.

Two recent examples of man-made disasters are the oil tanker named SS Torrey Canyon shipwrecked off the Cornish coast in 1967 a 120,000 tons vessel carrying crude oil and the Deep Water Horizon Oil spill in the Mexican Gulf 2010.

The Torrey Canyon disaster was significant to say the very least to the Southern British Coast line. Whilst the ship wreck of the Torrey Canyon happened down on the Cornish Coast, the pollution spill affected the whole of the English Channel, Thames Estuary and North Sea.

These waters were thick with crude oil affecting sea life and wild fowl with devastating consequences. I clearly recall fishing that year in 1967 where my fishing line was caked in crude oil and the smell considerable.

Many parts of the South Coast are significantly tidal, where the seas reseed considerably and in estuaries the tide in parts can go out for many miles. The crude oil deposits reached and contaminated every part of these estuary environments killing small marine life, molluscs and vertebrates.

The fish food chain from such devastation to these areas in no uncertain terms, was devastating and a common oversight surrounding the critical importance of coastal waters and estuaries, as these environments constitute main breeding grounds and nurseries for the rearing of young fish stock species.

Photograph Courtesy of Author Alastair R Agutter – Thames Estuary

In the Thames Estuary as one factual example as a result of the Torrey Canyon disaster, the Plaice flat fish stocks never really ever recovered. Throughout the Thames estuary there was an abundance of mussel beds, an attractive venue for Rag worm and harbour rag worm to thrive and a favourite staple food diet for Plaice especially and other flat fish.

The Torrey Canyon disaster did not only kill and pollute marine fish species but the food in these nursery breeding ground areas where both mussels and rag worm were devastated from the pollution. As a result of the food supply being affected, the Plaice as one example that survived changed their breeding habits and moved location to spawn and these new environments were more hostile for fry, such as the East of England Coast line waters of the North Sea and the South Coast areas in the English Channel, where more predators existed to feed on the fry of this species and others, therefore spawning levels were lower as a result of the water conditions and environment.

Spawning levels of fish are affected by conditions. If the water conditions are poor the likely hood of reproduction is more remote and the egg laying production amounts are also lower in number as a result of conditions and diet.

Rag worm and harbour rag worm in addition to mussels are high protein foods for example, the devastation to such food species of rag worm and mussels in nursery feeding grounds such as estuaries, forced fish to move away from these areas onto other breeding grounds. All wild life are affected by such man-made pollution events and so even the food supply species have to move location and this can take some considerable time, many years in fact for the few remaining species to recover and be sufficient in numbers.

On the East and South Coasts of England, for Plaice to breed in such environments successfully they require food for their fry to ensure fish stocks each year thrive, but food in these areas is far less abundant and where more young fry and adult species are competing for the food available.

The Plaice, Dab and Flounder stocks in the Thames Estuary never recovered properly as a result of such human industrial disasters. Many will claim however, fish stock levels are as a result of over fishing and this is true to some degree. But when there are major disasters, such as industrial man-made events it leads to species either becoming extinct or going in search of new safer breeding grounds and such environments are not always conducive for the species and where eventually the species population continues to dwindle.

Fish stock husbandry today is better than at any time in human history, but everyone has to participate and play by the rules. Very sadly many Countries still do not play by the rules and when in certain fishing grounds can have a long and lasting effect.

In the past common offenders on an industrial scale surrounding commercial fishing have been Russia, Japan, North Korea and Spain for over fishing using factory ships, virtually wiping out fish stocks to the point of extinction with regards to Tuna, Marlin, Shark, Halibut, Mackerel, Cod and other latter white fish species etc.

Photograph Courtesy of Author Alastair R Agutter – Small Inshore Fishing Boats

Today more measures are in place for commercial fishing, but again this all relates to man-made industrialization and the catch restrictions are often irrelevant, for under sized fish just like others naturally swim down a net when caught in an otter or beam trawl and these trap all size of fish in the belly end of the net known as the "Cod End" where there are countless numbers of reinforced nets and mesh to prevent the net from being torn open, when working on a sea bed and when being hauled aboard where the wait can be several tons.

In recent years many European Commercial Fishing boats and vessels have been de-commissioned where former owners were paid out a large lump sum to stop fishing and take up another career.

From man-made technology today in the way of sonar fish finders these can have a devastating effect on white fish stocks as these species often shoal such as Cod, Haddock, Whiting and Herrings etc.

When quotas are met by commercial fisherman (seasonal fish catch limits), if more of the same species are caught after reaching their quota, species have to be thrown back and returned to the sea to prevent the risk of fines or the impounding of a Skipper's vessel. All these fish returned are dead and another example of man-made industrialism serving to the detriment of the environment.

The continued onslaught to our seas from man-made commercial industrialism and pollution from factual examples given above mentioned. The future outcome for all life in our seas in the very near future is at a critical point to say the very least, or even more concerning we may have already gone beyond the point of return or recovery!

When an aquarium biological system breaks down the only real cure from nitrite and ammonia level build up is to literally replace most of the water and on a daily basis. This we certainly cannot do regarding the World's Seas, Estuaries and Rivers!

Climate Change Sea Levels are rising as a result of the Earth's atmosphere temperature increasing causing ice caps to melt.

In 1968 an eminent British Scientist Sir Wally Herbert expressed concerns of climate change by witnessing first-hand the melting of ice caps, one occasion upon reaching the North Pole after an epic journey described as one of the greatest human feats of all time.

Today in 2015 some 47 years later the stack of evidence surrounding sea levels rising and the melting of ice caps is over whelming. Coastal erosion has never been so great and the countless storms now witnessed around the world from our televisions destroying coastal defences and flooding coastal towns and regions is there for all to see!

The rise in sea levels is not the only disaster scenario that we as life forms on Earth are confronted with. As the Earth's atmosphere warms causing the melting of ice caps and sea levels to rise, more extreme and erratic weather conditions will continue to increase and these will be covered in more detail later regarding Climate Change to our lands.

The rise of sea levels also alter the chemistry of our oceans, estuaries and rivers around the world and with serious impact to all wild life inhabitants including the food chain to sustain all life on Earth.

If we take just one chemistry ingredient such as salt, from the rise of sea levels will therefore affect the specific gravity of salt content and this will in turn affect specie life forms from the most primitive of organisms to the most advanced predators in our seas namely sharks. Micro-organisms and enzymes at the beginning of the food chain are food nutrients and proteins for planktonic life forms of crustaceans, vertebrates and also fry larvae and eggs. Plankton as we know is a food source for shell fish species such as crab, krill, shrimps prawns etc. Plankton also provides food for all other young fish fry species and many advanced mammals of the Whale family.

The changes to the salt content in our oceans will and is threatening already many species, especially Micros-organisms and other more primitive marine life entities that cannot adapt, to the more advanced species that have greater tolerance variants in relation to specific gravity, salt content and regions in relation to temperature extremes but after a period of time these species will also be affected as the food chain begins to breakdown from the loss of smaller earlier food chain species mentioned.

The Gulf streams in our Seas are determined by the behaviour of our Jet Streams in the Earth's atmosphere. Changes to the Jet Streams have an impact on the Gulf Stream paths. These warms currents begin south of the equator and are drawn up by the cold waters north of the equator and these routes host countless thousands of marine life species and entities that migrate over season's regularly using the Gulf Streams as a natural food chain source of supply.

Photograph Courtesy of Author Alastair R Agutter – Canadian Brent Geese Feeding

When these Gulf Streams change as a result of Jet Stream changes in the Earth's atmosphere does not only impact above sea level regarding extreme storms leading to flooding and devastation. But the chemistry balance and reproduction cycles of marine life dependent on the Gulf Streams for food and reproduction cycles in our seas, rivers and estuaries.

Bird wild fowl migration is also affected by these changes to the weather in relation to Jet and Gulf Streams. Again nature works in concert, both bird and fish species work together and serve each other.

Canadian Brent Geese travel many thousands of miles around the world on migration paths visiting many estuaries that support sea grass, one of the geese's favourite foods. The tearing and eating of these grasses allows the sea grass to spread and in turn providing protection to fry fish species hatching in estuary nursery areas.

The recent dramatic and erratic changes surrounding the Jet and Gulf Streams flow paths have had a devastating impact on fish and bird migration. Nature demonstrates to us how fowl and fish work hand in hand regarding these historic journeys and as a result maintaining a balance in nature.

Fish stocks are not only affected by the change in Jet and Gulf Streams but also as a result of the aftermath surrounding heavy extensive rain in regions that then finds its way into rivers and estuaries. Where here there is a great impact on nursery breeding grounds for both marine and wild fowl from a large deluge of freshwater into a saltwater estuary environment.

Only certain fish and mammal species can live in sea and freshwater, most species cannot live in both and when in contact as a sea specie with a percentage of freshwater intensity can only withstand such conditions for a limited period only!

So if in a saltwater estuary environment where sea life species of fish have been spawning such as plaice, flounders, bass and dabs etc. The concentration of freshwater travelling into an estuary continuously due to river outage from regional flooding, this will not only affect the eggs and fish fry of the species laying the eggs, but also the food supply of micro-organisms including plankton, crustaceans and worms will move further down the estuary and in most instances, back out to sea where the salt content is specie life sustainable.

As a result of these climate change events to estuaries young fry will inevitably die on hatching or will not hatch at all if the eggs have already become fungal due to the pollution carried down and the change in water chemistry caused by freshwater, resulting years later in fish stocks again being devastated.

The continued flooding and excessive freshwater to these estuary regions will see greater loss of fish stocks and fishing will be poor in these areas and a rapid decline of other marine life species including crustaceans and vertebrates.

As sea levels increase and temperatures rise, oxygen content levels are greatly reduced. This allows algae to thrive, resulting with an increased presence of CO_2. In such conditions fish species tend to seek deeper cooler water to slow down their metabolism so they can endeavour to survive the conditions. This is very often seen in Lakes in relation to Brown and Rainbow Trout as two specie examples and where their feeding activities become more at night as they rest in the day in deep cooler pools.

In 2013 from the continued flooding from the River Thames into the Thames Estuary, I witnessed first-hand mullet gasping in numbers at the water's edge on beaches, no more than a few inches away from the shore line as they endeavoured to gasp for air and survive in the higher than normal concentration of freshwater found in the estuary and where as a result the salt content had been greatly reduced causing a slow poisoning to the fishes system and further exposed to vital organ breakdown and disease as often witnessed by Salmon when they return to rivers to spawn and eventually die after laying and fertilizing their eggs.

In 2011, 2012 and 2013 flooding in the United Kingdom was considerable as a result of climate change and the Thames Estuary, as just one marine region was adversely affected by the continued concentration of freshwater. In 2015 some two to three years later it was reported by Paul Gilson former commercial fisherman on the BBC Look East Programme, that Cockle Fisherman based at Leigh on Sea in Essex, who fish out in the Thames Estuary for Cockles and other shell fish, believed the great decline in Cockle catches was as a result of dredging in the Estuary! But this is not the case. The reproduction cycles of the cockles and the decline in numbers, is as a result of the intensive amounts of freshwater in the estuary over prior years from climate change affecting the specific hardness of the salt content and in turn affecting the reproduction of these shell fish species and others.

The Deep Water Horizon disaster in the Gulf of Mexico is another man-made industrial pollution example that has had a major impact on the region and especially to marine and wild fowl species. The contamination to shell fish, crustaceans and vertebrates has affected the livelihoods of many residents dependent upon the sea and estuaries for their living.

The recent natural disasters in the way of storms and flooding, plus the oil spill of Deepwater Horizon has taken its toll on the region and provides us with a prime example of the impact caused by both industrial pollution and climate change to which are both related to man's exploits. It is very unlikely with the continued erratic and extreme weather conditions, that the region will ever resemble any form of normality as was seemingly the case before these disasters.

In waters around the world today, just in fish terms there are over 33,000 different vertebrates and over 550 cartilaginous species. All these miracles of evolution and many with majestic colours especially around the Great Barrier Reef are now under threat from Climate Change and Industrial Pollution.

Climate Change does not only kill off species and I estimate within two decades over 50% of marine life we know today will become extinct. But also the creation of a new more hostile environment from the chemical changes for the emergence of new species and the further evolution of some existing species. However, such new species will invariably not be a friend to man and such species will be more resilient to the extreme weather conditions we are creating from climate change.

As I write this today in 2015 the Great Barrier Reef is under great threat from climate change regarding rising sea levels, increased temperatures and greater acid water conditions affecting the PH balance of the ecosystem caused by the rapid increase of $Co2$ in our atmosphere and in our seas. Certain crustacean species are now becoming more prevalent devouring everything in their path and the adjoining Oceans are seeing population explosions of aggressive and venomous jelly fish species devouring all.

As we are witnessing around the World today migration of people from climate change, this also happens with other life forms and especially evolved marine species.

Wild life creatures will always through necessity pursue food resources, so as environments change from global warning and climate change, so will the location and feeding habits of marine species.

In this past decade the population of Seals around the British Isles has greatly increased and many have even taken up residence in many estuaries as they seek out new food sources. But also comes with these movements are other predators that thrive on Seals especially as a stable food source. Many such species derive from the Shark Family namely Great White and Tiger Sharks.

UK Anglers and Commercial Fishermen know there are many resident members of the Shark family in the way of rays and sharks around the coastlines of Britain.

But the Great White and Tiger Sharks are not members that frequent British shores if at all. Sometimes Tiger Sharks do run up the English Channel and North Sea in the months of September and October on journeys towards better feeding grounds. But the Great White Shark is not one to have ever really frequented British shores if at all, but now as a result of Climate Change things are in fact changing and in the past few years there have become more sightings and understandably so, as Seals are a favourite regarding their stable diet.

24

I have been an angler and student of Marine life now for over five decades and as mentioned breeding some of the World's rarest marine species in captivity. The changes I have witnessed over the past five decades to our marine environment has been considerable and sadly a story of only decline.

I truly wished the evidence and events before me could be consigned to natural cycle events on Earth. But from deforestation near rivers to industrial pollution to our sees and now islands of plastic in the Pacific Ocean, such events and destruction are only as a result of man!

QUOTATION

"Every shoreline I walk with sounds of nature as the backdrop to such a majestic canvas of beauty, I find myself these days forever seeking forgiveness as I am shamed knowing to be man."

~ Alastair R Agutter

CLIMATE CHANGE TO OUR LANDS

In the annuals of history can be found many events and evidence surrounding civilizations where man has had a great impact to the environment and with devastating consequences.

As far back as the Mayan Civilization there is evidence showing from the high intensity of populations to certain regions of South America and with greater demand for more food to feed the inhabitants has resulted in flooding and disease.

As a result of the increase in population the Mayan's began a program of agriculture on the outer areas of new cities. Where they cleared forests and replaced these with fields for growing grain etc. As a result of such acts the structure of trees that wove a tapestry from their roots to hold river and stream banks saw land-slides, resulting in flooding to the regions and closely followed by disease namely dysentery.

In the United States of America the State of Oklahoma became a dust desert as a result of cotton farming on an industrial scale and where we saw human migration on a significant scale.

The City of Liverpool in Britain became known for disease due to the population growth and intensity in the Victoria age of industrialization.

The above are just some examples from our past where man from commercial industrialization in some form has had a significant and devastating impact to a region.

Today from climate change caused by man-made industrialization and high populations to regions we are now seeing more prevalent diseases in the form of Aids and Ebola and all related to the human impact on Earth.

Where in the past events surrounding man's industrial destruction and pollution was consigned to only regions, today with a human population of over 7 billion the impact is now a global one!

The continued cycles of man-made industrialization is no longer sustainable regarding the health of the planet. We are already seeing social unrest and conflict breaking out in many parts of the world as a result of a continued flawed method for society to function in the form of material wealth, namely money.

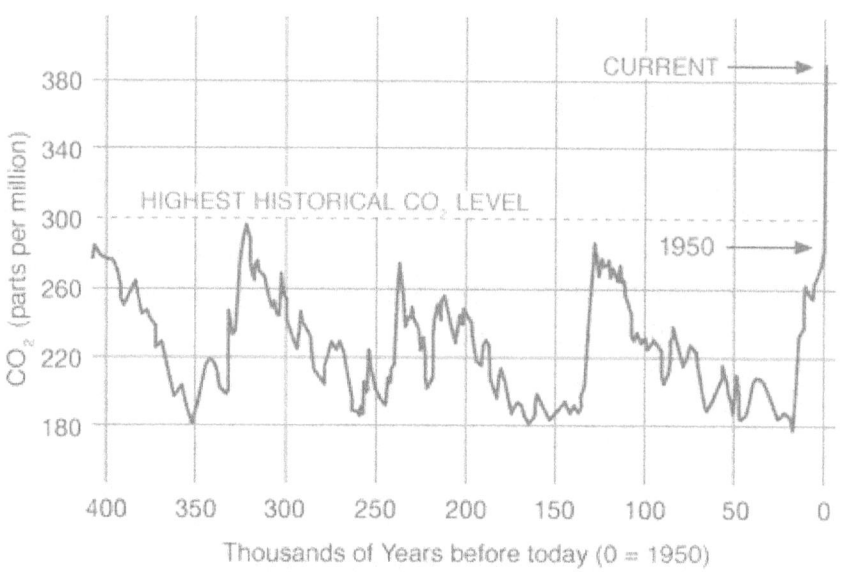

PROXY (INDIRECT) MEASUREMENTS

Data source: Reconstruction from ice cores.
Credit: NOAA

Photograph Courtesy of Author Alastair R Agutter and Data Courtesy of NASA
(www.nasa.gov)

Buying into such a flawed method for society to function are now new and emerging Countries such as India, China, Brazil, Nigeria and others. As a result what comes with such a spectre is more industrial pollution from the burning of fossil fuels and production of oil based products namely plastics!

I mentioned at the beginning of the book two specific numbers, these being 275 and 400. These as mentioned related to 275 ppm of Carbon Dioxide the beginning of all life on Earth and 400 ppm of Carbon Dioxide the current measurement in the Earth's atmosphere.

These numbers are factual scientific measurements as seen above in the picture and can no longer be ignored, but sadly Countries around the World are not addressing the crisis collectively that is the biggest threat to all life on Earth now! Still the child like mentality rules are being applied of self-interest and so no adult solutions are coming about, or are our great minds being allowed to advise and shape a better World for all.

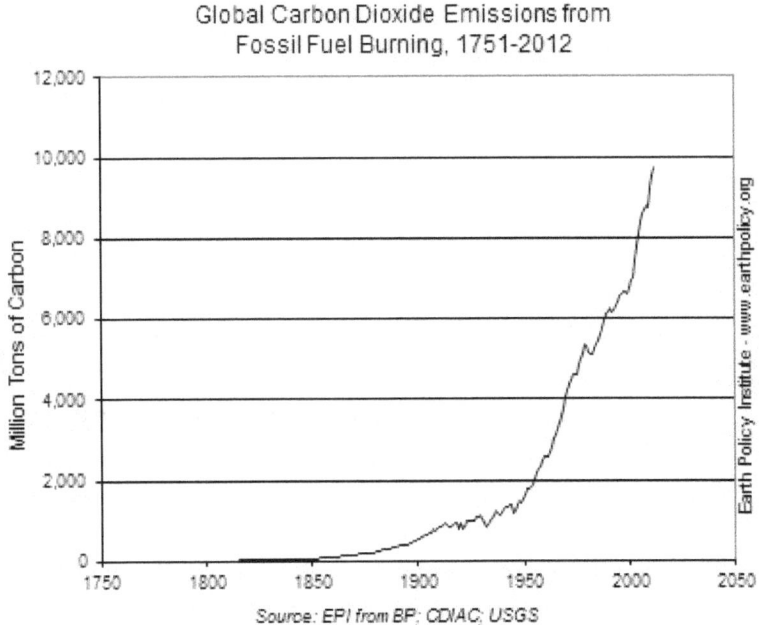

Photograph Courtesy of Author Alastair R Agutter and Data Courtesy of NASA (www.nasa.gov)

Still today sadly, the whole of life on Earth are at the mercy of many corrupt political officials and Corporate Empires who are responsible for this climate change crisis and where there only intent surrounds mutual superficial interests of money!

The failure to address this threat to all life on Earth we know as Climate Change from man-made industrialization, has only one outcome and this will eventually be all-out War!

Unless all of society changes and gives up on child like ideas of personal vanity, greed and self-indulgence that all relates to ignorance (education) at the end of the day there simply is no future!

This is a very depressing prospect yet such an outcome can be avoided if the interests of all citizens and life forms of the World are taken into consideration in the 21st century. We can no longer tolerate a few Kingdoms anymore who are also human rights violators and a threat of all life on Earth and are still allowed to produce more fossil fuels in Gulf States namely Saudi Arabia and Qatar. At the same time these rogue state dictatorships continue to be instrumental in conflict around the World to create instability to regions in the interests of sustaining their own personal status, wealth and greed.

Climate Change has been created from the above mentioned, Corporate Empires and Financial Institutions namely banks. Evidence is without question today surrounding the VW scandal over car emissions which amounts to criminal fraud, that these powerful entities have no regard for their customer audience or the environment. The reality surrounding Corporate Empires is that they will lie and cheat to the ends degree at the cost of all of human society and all other life on Earth.

The Financial Institutions continue to bank role such entities above and create a society of consumer credit and debt surrounding these corporate thugs' products and services. Such environments fuel hate and resentment leading to social unrest and extremism, all of which come with high environmental consequences, but also leads to mental health issues and other serious health conditions that will be covered later in the book.

In 2015 for the very first time meteorological institutions recorded weather speeds surrounding hurricanes and storms in excess of 200 mph around the World.

Such storms today have only grown in size surrounding Climate Change. The evidence for me with first-hand experience of storms dates back to 1987 in the United Kingdom where we endured a great storm. The wind speeds paled in comparison to today's storms, yet the storm of 1987 laid waste to trees that had been in the ground for many hundreds of years.

Great pines and formidable oaks were laid flat in rows like a folding pack of cards. Complete forest areas were laid bare for the first time in hundreds of years. The evidence of Climate Change is there for all of us to see when we know such trees graced England's landscape for hundreds of years.

Such events are not only natural cycles for if they were? Trees of over 800 years old in the way of Oaks simply would not exist, for they would have been destroyed from previous event cycles.

These continued events of extreme weather derive from man-made industrialism causing climate change for many of the World's great forests that absorbed Carbon Dioxide and other toxins were able to filter out such hazards as large forestation canopy nets located around the world.

But today as more forests are destroyed we are not learning from history where land masses become laid waste from extensive industrial farming and deforestation.

Eventually from a continued increase regarding a hostile and toxic atmosphere the very plant and tree life that protected us has now become a victim in many respects and affected by the continued climate changes, one recent factual example being the disease 'Ash Die Back' and where to date there is no solution or cure!

The destruction of the Amazon Rain Forest especially that gave us penicillin has in all probability seen the loss of other plant life forms that held the key for many cures to diseases we are experiencing today.

Socrates once said "what are you learning from this life" well for myself personally speaking, I have come to learn and understand that every living entity be it animal or plant life, all have a purpose and we cannot thrive without any of them on this fragile Earth we all know and call home.

In a well-balanced environment that is working just like an engine of a car, we should appreciate that all components are required to function.

There are many species on Earth that may appear to be pests or vermin, but I am sure they are in fact a friend. It is only our ignorance of failing to learn by studying the species to understand the great many secrets they hold and their importance in the grand scheme of things.

In the 1970's and 1980's there was a period of where acid rain was frequently discussed as a result of industrial pollution and then the discussion appeared to dwindle and eventually become grouped together and encapsulated in the climate change debate.

But acid rain has not simply disappeared it is more prevalent today than it has ever been and poses a serious threat to all humans and wild life on Earth. Many of today's ailments and health issues, are as a result of industrial pollution and will be covered later in the book surrounding climate change health and very possibly answering many searching questions.

In today's society many are seeking maintenance free easy solutions but all such come at a high price. One of the biggest contributors to flooding in urban and town areas is down to more parks and gardens being concreted over. Such actions greatly increase the risk of flooding and can be to the detriment of your health regarding airborne viruses where there is no longer a fully functioning biological filter system working at your home we commonly know as gardens.

Like our Seas that function from a biological system is also replicated on land, but very little is discussed and appreciated surrounding the functions of a garden and the wild life it houses. We have molluscs in gardens just as we do in the sea; we know them more as slugs and snails.

Our gardens believe or not are our first front line of protection surrounding our health and our immediate environment. Even our well-being by being stimulated by the vibrant colours of life has an important relevance covered later!

The grass in gardens work as a matted biological filter medium and a protector especially in times of excessive rain, where such a natural carpet serves to absorb the water like a sponge. The process of filtration as the water passes through the medium provides valuable trace elements for the successful growth of surrounding plants and also assists in the purification of the water as it works its way through the water table reducing the risk of prevalent bacterial development caused by stagnant water collection on concrete surfaces and finding its way into water tables and systems below.

Photograph Courtesy of Author Alastair R Agutter – More Gardens in Communities

Photograph Courtesy of Author Alastair R Agutter – Reducing Flooding in Communities

The cause of dysentery to many Victorian cities in Britain was as a result of impoverished areas where water and other waste found its way back into the freshwater table system.

Areas such as Canvey Island, Basildon and Southend on Sea in Essex, England are a few good examples today where from flooding caused by Climate Change have become regular victim towns suffering from excessive flooding as a result of creating concrete communities.

Many in society today believe climate change can only be resolved by politicians, but this is not necessarily the case. We can all do our bit by creating more grass and planted areas at our homes and in our communities.

We can also educate and remove incompetent Councils and Officials so there becomes more park and planted grass areas to soak up excessive rainfall and at the same time serve as micro biological systems in the community filtering out toxin and other hazardous chemicals caused by motoring emissions, namely diesel etc.

Another great concern and evidence found this year in 2015 relates to the devastation of insect life species in the environment caused essentially by industrial farming, commercial and domestic pesticides.

Photograph Courtesy of Author Alastair R Agutter

This year whilst writing a new book for the community to get more folk gardening, I discovered from the vegetables that I had been growing suffered as a result of the alarmingly low rate of Bees. Amongst the vegetables I planted, there were many other species of plants that could attract Bees for pollination, but I found on average more than 70% of all produce was not being pollinated. This I find to be of grave concern and put this further down to air pollution as covered more in the heath section that follows with my findings.

Photograph Courtesy of Author Alastair R Agutter

The weather conditions were good this year and the location of the vegetables were in an optimal location, so there was no excuse for crops to fail as they did grow well and in large numbers. But as I say, I have been alarmed to see so few Bees this year and this deeply disturbs me and another reason why we need to seriously look at all that we are doing! Especially surrounding commercial pesticides and where there is further greater evidence more than at any time, as to why we all need to grow and produce more organic foods and at home.

QUOTATION

"When dark clouds gather and on the horizon we see a storm coming, it is time to consider the most important things in life."

~ Alastair R Agutter

CLIMATE CHANGE TO HUMAN HEALTH

At the end of 2013, I collapsed from work exhaustion and in the new year of 2014 collapsed again! But on this second occasion it was from pain, where I was eventually diagnosed with having Rheumatoid Arthritis in the lower part of my spine. The pain was as a result of a spasm and from the damaged caused, I was unable to walk for over three months.

I have never been over weight and always been busy or athletic throughout my life. Upon evaluating the situation, where I recalled the colder months and when they approached how my breathing became erratic at times in the evening air especially and struggling to breath properly. To make absolutely sure of no chest infection, I took a course of amoxicillin penicillin and found again, as I recovered through the cold months there were many times especially in the evenings as the air cooled a difficulty in breathing once again.

One particular evening, I smelt in the Earth's atmosphere Sulphur Dioxide and seriously began to evaluate events leading up to my health problems and after. SO2 (Sulphur Dioxide) can often be found in the atmosphere from volcanic activity. But the more I investigated the events surrounding its presence in the atmosphere. SO2 when making contact with cloud formations it creates acid rain and knowing my condition surrounding bones and cartilage regarding my spine where the spasm occurred, I began coming up with a theory surrounding my condition and of the many other folk I kept hearing about who were suffering in the community around the same period of time, that especially related to cold and damp spells of weather.

In humans the average body water content is around 57%! For an adult male the body water content is normally around 55% and an adult female is normally around 60%. Now in view of the amount of Sulphur Dioxide in the atmosphere, along with the great increase in Carbon Dioxide, the tell-tale signs were all heading towards air pollution regarding health problems. I began to conclude my condition and others could be as a result of the toxins penetrating my body based on the fact it consists of 60% of water and when sulphur dioxide makes contact with water, it creates an acid water toxin.

Now such acids over a considerable period of time when the human body is exposed to it, will take its toll in the way of affecting body muscular tissues, cartilage and bones. In other words exposure leading to bone and cartilage erosion from acid rain caused by the body being made up of 60% on average of water. Such an atmosphere would also affect the lung tissue and breathing, causing respiratory ducts to inflame or close.

I returned to my Doctor and explained my theory, at the time he concluded it was feasible, but thus far to date had not heard of any known research in relation to my findings. But many months later, a friend informed me it was becoming a more common thought that conditions of stiffness, trouble breathing, asthmatic signs were as a result of the atmosphere and these were becoming frequent views of medical professionals in surgeries and clinics.

Such atmospheric conditions began to make sense regarding other ailments that are becoming more prominent in society surrounding degenerative diseases such as Alzheimer's and Dementia. For these such conditions relate to the deterioration of the brain cells in humans and this can be caused by the lack of pure oxygen reaching the brain, instead a slow poisoning process killing brain cells as a result of atmospheric air pollution namely sulphur dioxide SO_2 and carbon dioxide CO_2.

The more I evaluate my findings the more sense it makes, especially with the recent announcements by the medical profession that there will be a distinct increase in the number of case sufferers of Alzheimer's and dementia.

If the toxicity levels continue to rise such as Carbon Dioxide already in the 400 PPM range as researched and accurately recoded by NASA, this figure is heading towards twice the level for the beginning of all life on Earth, this being 275 PPM of Carbon Dioxide.

When the atmosphere warms up and the air rises, it therefore reduces the concentration of Carbon Dioxide at ground level and breathing becomes more normalized. However, when the air becomes humid from the heat the conditions resemble signs of hay fever, where eyes become irritated and sneezing becomes more frequent with runny noses. This again is signs of Carbon and Sulphur Dioxide penetration and irritation to the body. You may also experience signs of tingling, or wanting to scratch yourself in hot weather and this is caused from where a mild form of acid burning is occurring to the body, as the skin pours are more open when the body is overheating.

The reality is many medical conditions today are as a result of climate change and having a substantial impact on human and animal life populations. Another further sign of why society must change as we are now beginning to poison ourselves and all other life forms from industrial pollution, mainly caused by fossil fuel burning, resulting in a continued increase of Carbon Dioxide into the Earth's atmosphere.

In Britain the National Health Service can no longer sustain itself! This is as a result of demand on the system surrounding the rapid increase in mental health, degenerative diseases, arthritis, asthma, hay fever, flu symptoms, exhaustion, depression, dementia, Alzheimer's, obesity, back pain, spinal conditions and an ever increasing demand for joint replacements, all that I believe to be related to climate change, caused by industrial air pollution being the number one cause and factor!

Last but not least, I believe cancer is also attributed to environmental conditions and triggered more today from leading unhealthy stress related lifestyles, as cancer is a dormant form that resides in all of us.

As we continue to impact Earth from commercialized industrial pollution, many drugs today especially the penicillin class, are becoming less effective against virus diseases. This is as a result of climate change where a hostile polluted environment is seeing bacterial forms and diseases evolve further and becoming more resistant in pursuit of survival in an ever increasing hostile environment the human race has created.

Since my illness, I have studied in more detail commercial processed foods and found technology is creating an ever increasing chemically based solution in relation to food additives and ingredients, in simple terms creating a modern day diet of toxicity.

Many modern day packet processed foods contain, toxins, additives, emulsions and fillers etc. The list in fact goes on and they now all need to be consigned to the bin forever! For I have found from eating what I describe as proper wholesome food, in the way of home grown vegetables, fresh fish, natural butters, dairy products and meats. My health has rapidly improved with less pain and discomfort.

Today only when severe cold and damp weather arrives, am I really conscious of any discomfort. In milder and dry weather conditions, I do not suffer at all and find myself having more life to me and wanting to work on many more books and projects.

Unhealthy food diets in a modern day world namely packet processed foods, I believe to be not only unhealthy but do not provide the correct or sufficient proteins and vitamins the body needs, whereas healthy naturally grown foods do!

So to help the planet and ourselves today, we all need to grow more foods and plants for our healthy well-being and at the same time from healthy outdoor exercise helps the body, mind and soul.

QUOTATION

"Perhaps the secret to life is to understand simplistic perfection. A phenomena that surrounds us all each day in our natural world and by appreciating such beauty and wonder, can provide us with great fulfilment and enlightenment from the knowledge of learning."

~ Alastair R Agutter

CLIMATE CHANGE HUMAN INDUSTRIALIZATION

Just recently I had a conversation with a very close and dear friend who worked in Silicon Valley and for Morgan Stanley. We both concluded throughout the history of man be it tribal, man-made religion, political governance, to banking and products there was always a salesman, selling an idea!

Commercial Industrial Pollution comes in many guises and it is time to realise such a fact. Banking is a commercialized pollutant in society today along with debt and credit companies. As they create an environment where they sell the idea of credit to consumers all around the world so they can obtain this or that product. Many are not required at all in our lives, but the World has bought into such superficial ideals and now as in the past this comes at great cost.

Throughout the ages and history of man from empires, states or movements, namely man-made religion has resulted in conflict. Then after such an event, a rebuilding process begins yet again. Not once, twice or three times, but continuously throughout the history of man!

Such activity should be evidence enough that the repetitive path taken surrounding money and consumerism in many forms is both destructive and futile!

With the troubles in the world today and the power of Atomic weaponry, it could well be the case that the next conflict we will be one of permanent demise for the human race! A Nuclear war will produce no winners only a holocaust beyond human comprehension and when Robert Oppenheimer said he had become death and that was not strictly true in fairness to him. But what he had created was a German Nazi concept known as the final solution!

Today's society of consumerism is a car crash waiting to happen, the banking system is both morally bankrupt and flawed and corporate capitalism is a polite way of saying liars and cheats to make money at any cost.

Today's society believe they have a God given right to exist and regard all other miracles of evolution now struggling as fair game to exploit, or perceived to be pests or vermin, as recently described by some mindless idiot when referring to Sea Gulls taking chips from humans, as a result of their feeding grounds being totally destroyed and wiped out by the rapid deterioration of our seas, estuaries and rivers.

The human migration underway at this time, that we are all a witness too and now leading into Europe, where we see a trail of rubbish and litter, such events aptly depict the mess we as a species are in, where there is no real regard for our planet, or our wild life. Only a continued drive of destruction surrounding the desire for money, to buy more consumer products and services, such an environment regarding the human condition that has become one of the biggest threats surrounding mental health and will never come with a healthy outcome unless change is made.

The reality is behind the smiles of the selfies like a scene out of a Martha Stewart presentation there is either emptiness, lack of self-esteem, confusion, anxiety or addiction with the belief that all would be right in their lives if they all had money.

The reality is the planet is dying and the most precious commodity is life itself!

Family loved ones and relationships are important, the relationship with our Earth and having a regard to comprehend, love and respect it and to connect with Mother Nature can provide the fulfilment of a hundred life times.

There are countless millions of life forms here on Earth and all evolving seeking simplistic perfection and there to study and learn from as to how the entire cosmos and universe(s) function.

Yet we as a species fulfil our world with desires of superficial material wealth that will last an hour, a day, a year or maybe a little longer. The only lasting tangible record one can keep is a memory nothing more and nothing less. Materialism cannot offer eternity only a brief temporary moment of self-gratification and this very sadly is the shallow concept of human society today!

The reality is we as a species and a planet are in trouble, our so called trusted brands and institutions are now becoming exposed for what they truly are like FIFA and VW. In truth they do not give a dam about anyone or anything, only themselves like a band of tribal thugs.

If the human race is to survive and as custodians, being the primate specie has a duty of care towards all other life forms things must change.

There is no longer a place in modern day society in the 21st century for self-indulgent greed and self-interest. The human race as a specie can only evolve beyond this point collectively in the interests of all other life forms that we share this planet with. For they all have rights and have just as much right to exist as we the human race!

There are enough great minds in this world today across states and borders to create new clean energy and to direct us in the right direction as an advancing evolving species, so eventually and only possibly by studying the miracles of life here on Earth and understanding how they function and evolve will be the key to unlocking the cosmos.

Fossil fuels in the form of oil and gas along with the vast production of consumer goods and products in the way of plastics are killing the planet. The rise in Carbon Dioxide levels now today is a most dangerous and life threatening point reached and has not risen over centuries, but surged in the past decade, just the last 10 years!

The days of Victorian industrialization is over and the continued burning and production of plastic superficial goods must now stop.

It is now time for change and I leave you with words said by Pope Francis to the United Nations Assembly of Political Leaders, all who are connected to the Industrial Commercialized World they have bought into!

I have no issue with man-made religion such as the Catholic Church if it makes better human beings through love, understanding and enlightenment. The only time I do have an issue with man-made religion and falsehoods is when it is used as a tool to kill and hate that defiles the name of the Divine!

QUOTATION

"A selfish and boundless thirst for power and material prosperity leads both to the misuse of available natural resources and to the exclusion of the weak and disadvantaged."

~ Pope Francis

QUOTATION

"At the time of your reckoning and where you reflect and evaluate the life you have lived. Only memories will be a commodity you can take with you, for all else is just material and of no value."

~ Alastair R Agutter

CLIMATE CHANGE TECHNOLOGY, TRADES AND CRAFTS CAN EMBRACE CHANGE

Just recently Mark Carney Canadian born Governor to the Bank of England alluded to the future and announced in a very subtle way the game is up, regarding banking and finance as we know it. That can only be described as a commercial industrialized pollution in the 21st century. He mentioned bit coins as an example regarding a future method of trading throughout the World.

I describe Mark Carney as sharp and nobody's fool type of man, he is also an astute visionary who latterly thinks, but keeps many of these thoughts to himself, even though he can see on the horizon a possible viable outcome to serve the interests of all and humanity.

As I have frequently mentioned we do have to change and time is running out, if it has not already done so. The production of fossil fuel consumption and consumerism products relating to plastics has to end.

Society needs to begin addressing in an adult grown up way the damage thus far done to the environment and this begins with developing practical innovative organic products.

On a recent visit, a young family friend had just recently completed her degree in creative design. I shared with her some of my thoughts and also showed her at the same time failed crops from a lack of Bees pollinating and said to her "it is now her generation to try and put right the mess today's humans have created!"

We need to develop in the World as of yesterday a massive tree planting programme, the likes of which never seen before and so we can produce again environmentally friendly organic goods in timber and many other products in metals that last a life time and not just last five minutes that end up on a land fill or on the continent of Africa and India where we can see a mountain of refuse. In simple terms making the planet into a dustbin!

I recall as a child some of the fabulous toys made and crafted out of wood. Furniture and Tables crafted masterfully that would last a life time. These are the products we need to produce again and with the sensible mass planting and management of trees, so we can again begin to restore the canopy of nature that filters out many toxins and chemicals in the atmosphere that threaten all of life.

Travel and transportation is possible today by taking a quantum leap and exercising our innovation skills, to produce transportation networks that can function and operate from solar energy and magnetic gravity. If the Romans could build Aqua Ducts that could travel across borders and states several hundred and thousands of miles by using gravity, there is no reason why we cannot today build rail networks in the same way.

Solar energy flights have just recently happened and there is no reason why we cannot evolve and develop these as a commercial means of air travel for all.

Mobile and other communications technology today can now be solar, I recall friends at Netgear Bay Networks informing me they are developing solar hubs and switches.

If we can flat pack the web today by using optimization technology as covered in my book notes publication titled "Creating the New Internet Super Highway" there is no reason why we cannot adopt such a concept across many technology fields in the interests of all and even become more creative and imaginative, by developing shell products so we can upgrade without having to completely replace an item.

We can save the planet and ensure the future of the human race and all life forms, if humans begin to reach their full potential and be big enough to collectively work together in the interests of all. Starting with the re-structuring of the United Nations so such an institution has a purpose for all of humankind. Then if desired we can all go and start growing crops in the African basket for the World and all go and plant trees in Indonesia,

Malaya and Brazil to start recovering the Rain Forests Earth so desperately needs and depends upon.

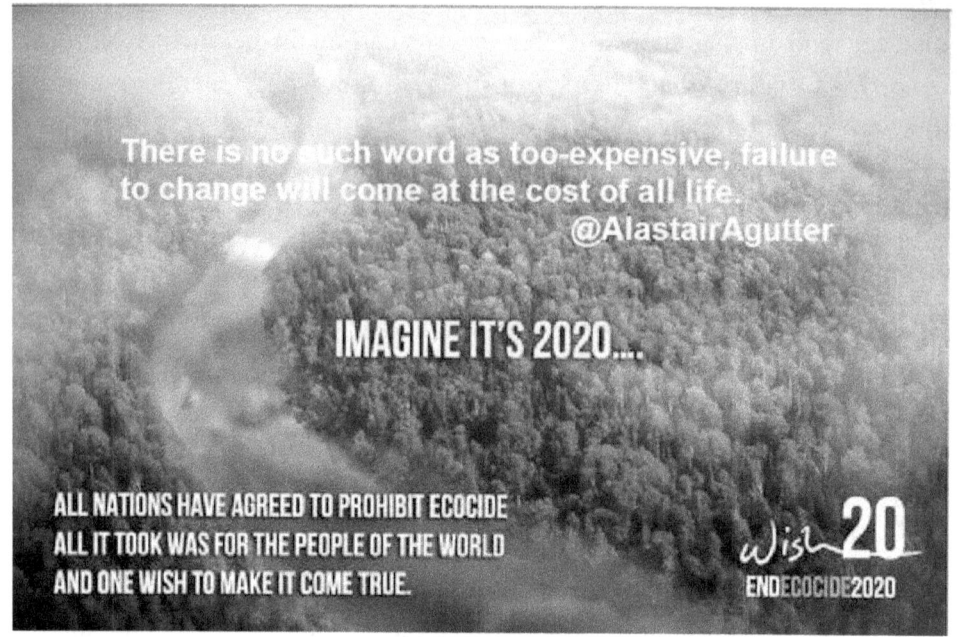

There is no such word as too-expensive, failure to change will come at the cost of all life.
@AlastairAgutter

IMAGINE IT'S 2020....

ALL NATIONS HAVE AGREED TO PROHIBIT ECOCIDE
ALL IT TOOK WAS FOR THE PEOPLE OF THE WORLD
AND ONE WISH TO MAKE IT COME TRUE.

Wish20
ENDECOCIDE2020

Photograph Courtesy of Author Alastair R Agutter

Develop and build green homes where citizens can grow healthy organic free crops at the same time creating life and beauty, that act as micro biological filtration systems in communities and preserving all other life entities that are miracles of evolution who share such an environment.

The technology is there now to develop sustainable battery cells storage for energy in homes, fossils fuels could end tomorrow with the will of all and boycotting the suppliers of death (Oil Companies).

Dispense with plastic packaging products and produce again environmentally friendly hessian, paper bags and containers.

Use technology in a responsible and balanced way and not become dependent on technology as warned by Albert Einstein.

There is great opportunity if the World grows up and embraces change that will serve the interests of all and not the few!

CLIMATE CHANGE EMINENT QUOTATIONS

Climate change poses at least as big a threat to the world as war, the new UN secretary general, Ban Ki-moon, warns.

"We shall need a new way of thinking if humanity is to survive"

~ Albert Einstein 1954

Global warming is not just the greatest environmental challenge facing our planet — it is one of our greatest challenges of any kind.

~ President Barack Obama

The Science is in. The facts are there that we have created, man has, a self-inflicted wound that man has created through global warming.

~ Arnold Schwarzenegger

"The warnings about global warming have been extremely clear for a long time. We are facing a global climate crisis. It is deepening. We are entering a period of consequences."

~ Al Gore. Former Vice President (USA) and Nobel Peace Prize Winner

"Any harm done to the environment, therefore, is harm done to humanity. Human Being's, are not authorized to abuse it, much less to destroy it."

~ Pope Francis

"Today, more than ever before, life must be characterized by a sense of Universal responsibility, not only nation to nation and human to human, but also human to other forms of life."

~ Dalai Lama

"I once asked our Divine Father of all, what is the point of trying to warn and help the Human Race if they do not listen.

He replied at least you have tried."

~ Alastair R Agutter

ACKNOWLEDMENTS

"Sir Isaac Newton, Albert Einstein, Dalai Lama, Barack Obama, Pope Francis, Al Gore, Arnold Schwarzenegger, Ban Ki-moon, NASA, Google, Wikipedia, Natural History Museum, Science Museum, London Zoological Society, National Geographic and United Nations."

QUOTATION

"To some my ideas may seem a little naïve in such a troubled and uncertain world, but such wishes are the only beacon of hope we have."

~ Alastair R Agutter

AUTHOR'S WEB SITE

www.alastairagutter.com

www.ingramcontent.com/pod-product-compliance
Lightning Source LLC
Chambersburg PA
CBHW071003180526
45168CB00003B/1273